This notebook belongs to:

Absolute Value

Absolute value means how far a number is from zero.

$|-3|$ and $|3|$ are both 3 units away from zero.

So, $|-3| = 3$ and $|3| = 3$.

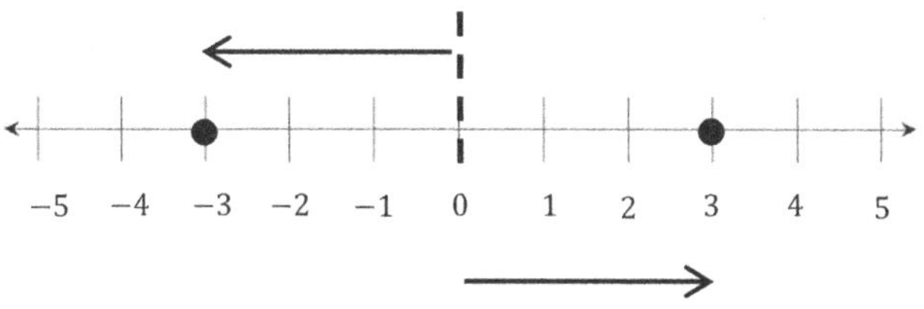

Absolute Value Properties

- $|a| \geq 0$ because $|x|$ can never be less than zero.
- $|a \times b| = |a| \times |b|$
- $|u| = a$ is the same as $u = \pm a$. This is key for solving most absolute value questions.

$<$ or $\leq$ Absolute Value Inequalities

$|x| < 2$ means the distance from x to zero must be less than 2.

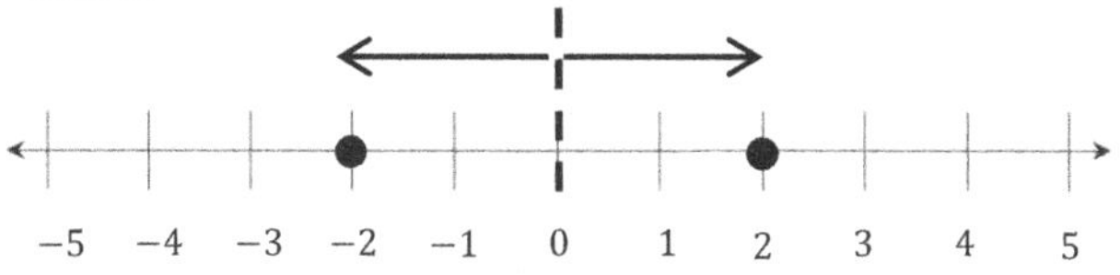

$-2 < x < 2$

Solving $<$ or $\leq$ Absolute Value Inequalities

- Solve $|2x - 6| < 4$

Rewrite as: $-4 < 2x - 6 < 4$

Add 6. $2 < 2x < 10$

Divide by 2. $1 < x < 5$

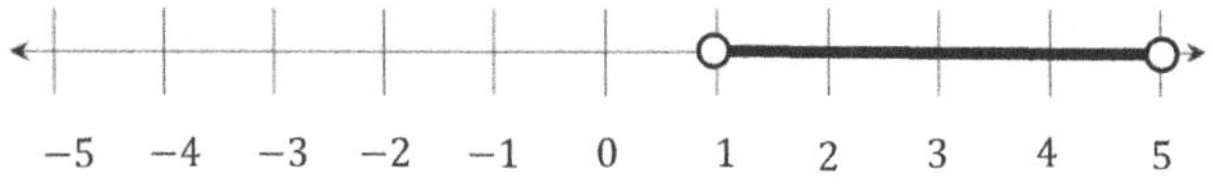

Solving Absolute Value Equations

- Solve $|x + 3| = 1$

Since $|u| = a$ is the same as $u = \pm a$ then,

$|x + 3| = 1$ is the same as $x + 3 = \pm 1$

Which has two solutions:

$x + 3 = -1 \qquad x + 3 = 1$

$x = -4 \qquad x = -2$

$>$ or $\geq$ Absolute Value Inequalities

This is different as there are two separate intervals.

- Solve $|x| > 2$

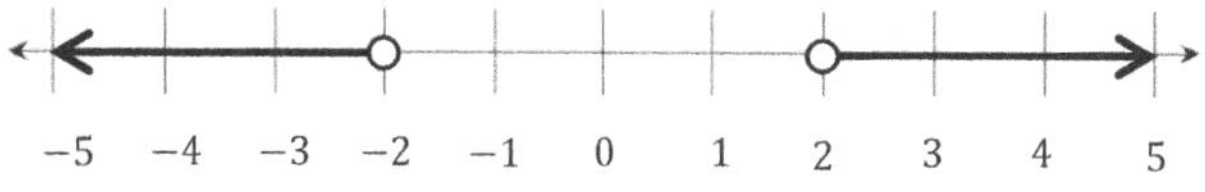

$x < -2$ or $x > 2$

Solving $>$ or $\geq$ Absolute Value Inequalities

- Solve $|2x - 4| \geq 4$

Rewrite as: $2x - 4 \leq -4$ or $2x - 4 \geq 4$

Add 4. $2x \leq 0$ or $2x \geq 8$

Divide by 2. $x \leq 0$ or $x \geq 4$

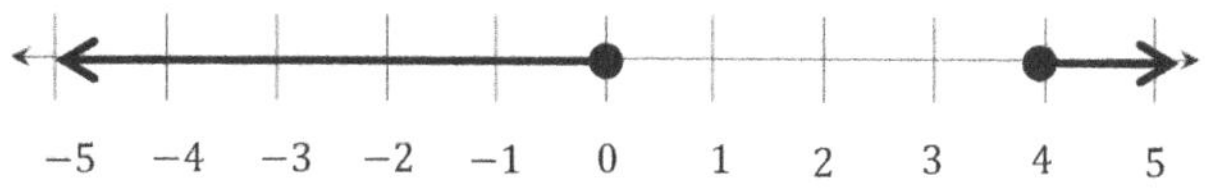

Made in the USA
Monee, IL
08 November 2026

68002531R00079